AF808033

# Move It!

by Mark Fara

PROPERTY OF
JOINT SCHOOL DISTRICT #251
JEFFERSON COUNTY
NO. ____________

Harcourt
SCHOOL PUBLISHERS

Orlando Austin New York San Diego Toronto London

Visit *The Learning Site!*
www.harcourtschool.com

There are more kinds of animals than anyone can name. They are active in different ways. Some of them walk and run. Some of them swim. Some crawl. Some even fly through the air! Sometimes an animal can move in more than one way.

Monkeys use their hands to help themselves move. They walk on both their hands and feet. If you watch one in a zoo, you may see it climb a tree.

Dogs and cats walk and run on four legs. Dogs can not climb, but cats can. A cat can climb a tree so fast that a dog can not catch it.

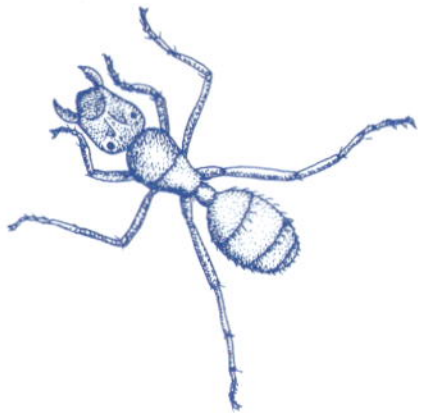

Insects have six legs. Many insects move by crawling. An ant can crawl up your leg. Eww!

An octopus lives underwater. It has even more legs than an insect. It has eight legs! But it does not use its legs to swim. It swims by sucking water into its body. Then it forces the water out very fast. The moving water pushes the octopus where it wants to go.

Fish live underwater, too. They have fins instead of arms and legs. They move their fins and wiggle their tails to swim where they want to go.

Many animals can move through the water by using their arms and legs. None of them swims as well as most fish.

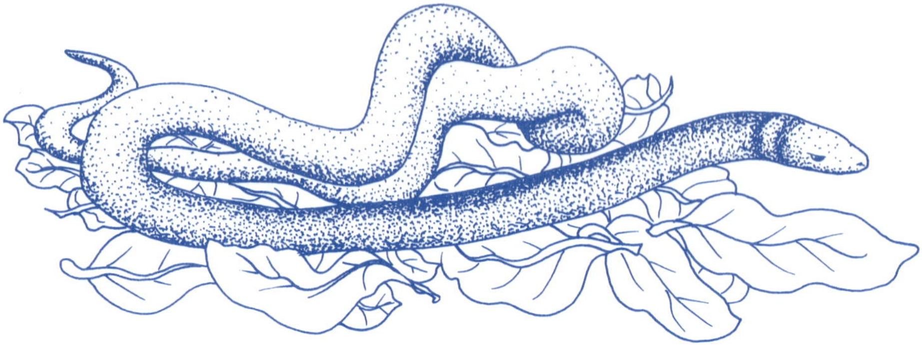

Snakes do not have arms or legs. They do not have fins, either. They wiggle their long, skinny bodies to move around. The way they wiggle on the ground is called slithering.

Kangaroos hop by using their two strong back legs. If you tried to catch a kangaroo, you would get very tired. Adult kangaroos can hop as fast as a car travels!

Birds have two thin legs. They have wings instead of arms. By moving their wings, birds can fly through the air.

Birds fly to look for food. They fly to carry food to their young. Some birds fly south for the winter because it is warmer there.

So the next time you see an animal, think about how it moves from place to place.

Does it crawl like an ant?

Does it swim like a fish?

Does it slither like a snake?

Does it hop like a kangaroo?

Does it fly like a bird?

Or does it walk and run, like you?